## creatures of the sea

# Sea Urchins

**Other titles in the series:**

Beluga Whales

The Crab

Dolphins

Flying Fish

Humpback Whales

Killer Whales

Lobsters

Manatees

Moray Eels

The Octopus

Rays

Sea Anemones

Sea Horses

Sea Otters

Sea Stars

Sea Turtles

Sponges

Squid

The Walrus

The Whale Shark

# creatures of the sea

# Sea Urchins

### Kris Hirschmann

**KIDHAVEN PRESS**
*An imprint of Thomson Gale, a part of The Thomson Corporation*

Detroit • New York • San Francisco • San Diego • New Haven, Conn. • Waterville, Maine • London • Munich

© 2005 Thomson Gale, a part of The Thomson Corporation.

Thomson, Star Logo and KidHaven Press are trademarks and Gale is a registered trademark used herein under license.

*For more information, contact*
KidHaven Press
27500 Drake Rd.
Farmington Hills, MI 48331–3535
Or you can visit our Internet site at http://www.gale.com

**ALL RIGHTS RESERVED.**
No part of this work covered by the copyright hereon may be reproduced or used in any form or by any means—graphic, electronic, or mechanical, including photocopying, recording, taping, Web distribution or information storage retrieval systems—without the written permission of the publisher.

---

**LIBRARY OF CONGRESS CATALOGING-IN-PUBLICATION DATA**

Hirschmann, Kris, 1967–
  Sea urchins / By Kris Hirschmann.
    p. cm. — (Creatures of the sea)
Includes bibliographical references and index.
  ISBN 0-7377-3012-9 (hard cover : alk. paper)
 1. Sea urchins—Juvenile literature. I. Title.
  QL384.E2H57 2005
  593.9'5—dc22

2004027627

---

Printed in the United States of America

# Table of Contents

**Introduction**
Learning from Sea Urchins   6

**Chapter 1**
Sea Porcupines   9

**Chapter 2**
The Sea Urchin Life Cycle   18

**Chapter 3**
Eating and Defense   29

Glossary   41

For Further Exploration   43

Index   45

Picture Credits   47

About the Author   48

## introduction

# Learning from Sea Urchins

In oceans around the world, sea urchins are a common sight—but not always a welcome one. Many careless waders, snorkelers, and scuba divers have brushed up against sea urchins only to be pierced by these animals' sharp **spines**. Some sea urchins can even poke their spines right through the tough material of a wet suit. The injuries caused by sea urchins are very painful, so people tend to fear and avoid these spiky creatures.

Most people who dislike sea urchins do not realize one important fact: Sea urchins are extremely sensitive to poor water quality and often die long before other creatures show signs of stress. This means an area with lots of urchins is guaranteed to

# Learning from Sea Urchins 7

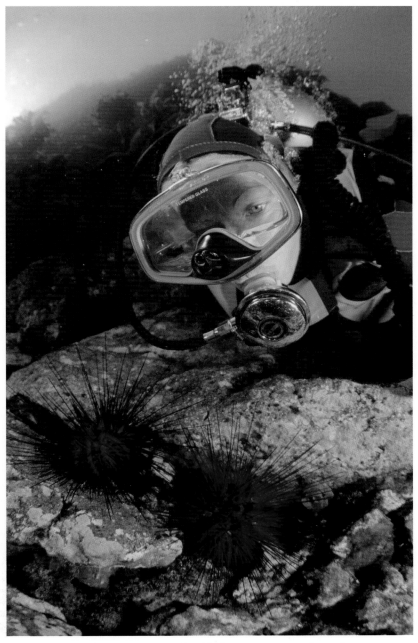

*A diver exploring a reef off the New Zealand coast keeps a safe distance from a pair of spiky sea urchins.*

be clean, healthy, and stable. For this reason, sea urchins should actually be a welcome sight to anyone who loves the ocean.

Partly because of their sensitivity to pollution, sea urchins are very useful in laboratory studies. Scientists gather urchin eggs and sperm and expose them to pollutants, as well as different temperatures, low oxygen levels, and other conditions. Then they look at the eggs and sperm under a microscope to see how they behave. Studies such as these teach scientists a lot, not just about sea urchins but also about humans. This is because sea urchin eggs and sperm are very similar to human eggs and sperm. If something affects sea urchin reproduction, there is a chance that it will affect human reproduction as well.

Sea urchin studies may be important to people in the future as more and more chemicals enter the world's water and food supplies. They are also interesting for their own sake. Like all ocean animals, sea urchins have many unique traits. Learning about these traits adds to our knowledge of life on Earth. This knowledge makes the natural world an altogether richer and more interesting place not only for those who visit the oceans and their bustling sea urchin communities but also for people everywhere.

# Chapter 1

# Sea Porcupines

In practically every region of the world, parts of the seafloor are covered with blankets of living, moving spikes. These spikes, also called spines, belong to creatures called sea urchins. Because of their prickly outsides, sea urchins are sometimes called sea porcupines or pincushions of the sea.

There are about 700 species of sea urchins. All of them belong to the **echinoderm** group, which also includes sea stars, sea lilies, brittle stars, and sea cucumbers. Within this group, sea urchins belong to the scientific class **Echinoidea**. This class also includes sand dollars and heart urchins, the sea urchin's closest relatives.

Sea urchins are strictly saltwater animals. They are never found in fresh or brackish (partly salty) water.

*There are nearly 700 species of sea urchins found throughout the world, including these giant red urchins.*

This is one of very few limitations on where sea urchins can live. These creatures are found in salty waters of every temperature, from the tropical seas near the equator to the freezing regions near Earth's poles. They also live at every depth, from shallow coastal waters to plunging ocean trenches. They are especially common in deep waters, sometimes gathering in vast herds along flat stretches of ocean bottom. Scientists believe that sea urchins may be the most common **macroscopic** (visible to the naked eye) animals on the deep-sea floor.

Although sea urchins can and do live in tidal waters, they are not as common in these areas. Sea urchins cannot live out of the water. They overheat and dry out if they are exposed to the sun for more than a few minutes. So they do not like places where the receding tide leaves them exposed. Sea urchins will, however, live in rocky holes along the shoreline that stay filled with pools of water during low tide. **Tidal pools** are good places for a person to see sea urchins without having to enter the water.

Wherever they are found, sea urchins usually settle on hard surfaces, such as rocks and coral reefs. Some deepwater species, however, live on muddy surfaces. This is an important adaptation in a region where mud flats may stretch for vast distances.

## Covered with Spines

The sea urchin's most distinctive feature is its spine-covered body. A sea urchin can have hundreds of spines, poking out in all directions except straight down. The longest, most visible spines are

*The slate pencil sea urchin is easily identified by its thick, stubby spines.*

called **primary spines**. Most sea urchins also have much shorter **secondary spines** that fill in the spaces between the longer spikes. Both types of spines help sea urchins to move, dig, and defend themselves.

Although all sea urchins have spines, the size and shape of the spines vary from species to species. Some sea urchins, including the slate pencil urchin and the ten-lined urchin, have thick, stubby spines. Others, including the spiny urchin and the banded urchin, have thin, needlelike spines. One species, the shingle urchin, has flattened spines that cover the body like plates or scales. This natural armor protects the shallow-water shingle urchin from battering waves.

The length of their spines is another way to tell types of sea urchins apart. The collector urchin and the West Indian sea egg, for example, are covered with short spines about ½ inch (1.2cm) long. The spines of other sea urchins may be much longer. The magnificent urchin, for example, has spines that are more than 4 inches (10cm) long. The appropriately named long-spined urchin has hundreds of needle-sharp spines that measure more than 12 inches (30cm) from end to end.

Spine color is yet another way to identify sea urchins. These creatures' spines can be brown, red, white, black, purple, green, yellow, pink, or almost any other color or combination imaginable. The spines of the Caribbean reef urchin, for example,

*The thin, needlelike spines of the long-spined urchin can grow to more than a foot in length.*

have purple tips, greenish shafts, and white rings around the base. Some sea urchins even have banded, spotted, or mottled spines. One example is the Thomas's urchin, which has cream-colored spines with brown bands.

## A Hard Skeleton

A sea urchin's spines are attached to a hard skeleton called the **test**. All sea urchins have the same basic test shape. It looks like a ball that is slightly flattened on the bottom. Test size varies from species to species. It ranges from just a fraction of an inch (a few millimeters) to more than 12 inches (30cm) across. The test is made from ten bony plates that are shaped like the

*The test, or skeleton, of a sea urchin shows the holes where its spines and tube feet were once attached. Its mouth is also visible.*

colorful segments of a beach ball. In most species these plates are joined along their edges. In some deepwater species, however, the plates are held together only by skin. When these sea urchins are taken out of the water, they slowly collapse as their skeletons fall apart.

A sea urchin's test is not a smooth, featureless ball. It has holes on its upper side that allow waste products, eggs, and sperm to leave the body. The test also has a hole that allows water to enter the body. A large gap in the center of the test's lower side contains the sea urchin's mouth and teeth. The most obvious feature on the test is a covering of little round bumps, each of which marks the place where a spine is attached. The bumps run in neat rows down the sides of the test.

Most of the test's features are not visible while a sea urchin is alive. They are hidden by the creature's spines and by a thin layer of skin that may be many different colors, depending on the species. It is not until the sea urchin dies and its living parts rot away that the beautiful test beneath can be seen.

# Tube Feet

Also visible on an exposed test are many small holes that dot the areas between the bumps. In life, these holes allow organs called **tube feet** to stick out of the sea urchin's body. The tube feet are long, thin, and fleshy. They end in tiny suckers that can cling to things. The sea urchin uses its tube feet for walking, climbing, and holding on to hard surfaces.

Tube feet have no bones or other hard material to keep them rigid. Therefore the sea urchin controls its tube feet with water pressure. An urchin sucks in seawater through the hole on top of its test. It pumps the water into a tube foot to stiffen it, then uses muscles in the foot's wall to direct the foot's movement. The muscles can also push water out of the foot to make it

*This urchin uses its tube feet to walk along a reef off the coast of Scotland.*

soft and flexible again. By alternately filling, moving, and emptying different tube feet, a sea urchin can travel anywhere it wants to go. It may also use its spines to give itself an extra push.

Tube feet are used for other functions besides walking. Some tube feet are designed to take in oxygen. These feet serve as the sea urchin's breathing organs. Other tube feet end in tiny three-pronged pinchers called **pedicellariae**. The sea urchin uses these pinchers to clean itself and to protect itself from predators. In most species, the pedicellariae are too small to be seen. In some species, including the flower urchin, the pedicellariae are large and clearly visible.

## Getting Around

A sea urchin is **radially symmetrical**, which means its body parts spread out evenly from a central point. In other words, a sea urchin has a top and a bottom but no front or back. For this reason, a sea urchin can travel in any direction without turning its body.

Sea urchins depend on a few basic senses. They do not have eyes, but they do have a nerve ring near the bottom of their bodies that has sensors that detect light. These sensors help a sea urchin to tell the difference between day and night and between open and closed spaces. Light information may also help sea urchins to defend themselves when predators approach. Finally, most sea urchins have sensors that detect gravity. This allows them to tell

# Sea Porcupines

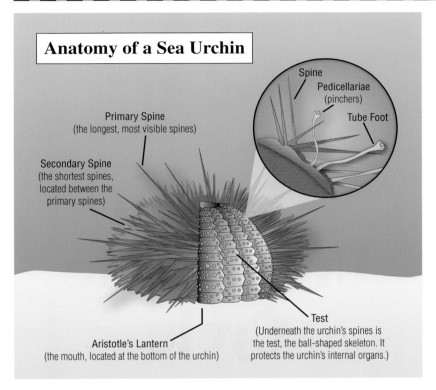

**Anatomy of a Sea Urchin**

- Spine
- Pedicellariae (pinchers)
- Tube Foot
- Primary Spine (the longest, most visible spines)
- Secondary Spine (the shortest spines, located between the primary spines)
- Aristotle's Lantern (the mouth, located at the bottom of the urchin)
- Test (Underneath the urchin's spines is the test, the ball-shaped skeleton. It protects the urchin's internal organs.)

the difference between up and down, so they can keep the top of their bodies pointing upward as they travel.

Sensory information is collected and sent throughout the sea urchin's body without the help of a brain. The lack of a brain proves that sea urchins are very simple animals indeed. Luckily, sea urchins do not need a brain to stay alive. Scientists have found sea urchin fossils that show that these animals have been on Earth for nearly 500 million years—much longer than the oldest dinosaurs. Sea urchins may not have brains, but, unquestionably, they have all the tools they need to survive.

RODNEY THOMPSON MIDDLE SCHOOL LIBRARY

# Chapter 2

# The Sea Urchin Life Cycle

Scientists do not know how long most types of sea urchins live. Based on the species that have been studied, however, it seems that their life spans vary widely. Some smaller sea urchins may live just a year or so, while larger species may live for several decades. At least one species—the red sea urchin—can survive even longer than this. Recent studies show that some red sea urchins are more than 200 years old.

Even at extreme ages, sea urchins show very few signs of aging. This means, in part, that a healthy sea urchin can continue to reproduce throughout its lifetime. By helping to create new sea urchins, an individual plays an important role in the ocean community.

## Spawning Season

Sea urchins reproduce by **spawning**, or releasing eggs and sperm into the water. Spawning happens at different times of the year for different species. Many types of sea urchins spawn in spring or early summer, when the ocean waters begin to warm up. Others do not seem to need warm temperatures. They spawn during even the coldest parts of the year.

When the time to spawn approaches, female sea urchins start to produce eggs inside their bodies. At the same time, males produce sperm. Eggs and

*A West Indian sea egg in the Caribbean Sea releases clumps of eggs into the water.*

*When urchins like these in the Pacific Ocean are ready to spawn, they release a chemical signal to other urchins.*

sperm are produced and stored inside internal organs called **gonads**. A sea urchin has five gonads that look like hanging clusters of berries. These clusters are connected to the inner wall of the test by thin strands of flesh. They are attached by thin tubes to small holes on the test's upper surface.

A sea urchin's gonads are large compared to the rest of its body. Even when it is not spawning season, these organs fill much of the space inside the test. They become even bigger as they fill with eggs or sperm, swelling to fill nearly the entire test cavity. They also change color. In a healthy, well-fed sea urchin, the full gonads appear bright yellow or orange.

When it is almost time to empty the gonads, all the sea urchins in an area start to release chemicals from their bodies. These chemicals float through the water to other sea urchins, confirming that the moment of spawning is approaching. Excited by the chemical messages, all the sea urchins rush to finish their egg and sperm production.

Finally the right moment arrives, and the sea urchins begin to release their eggs and sperm into the water. It will take one to two days for all the sea urchins in the area to spawn. During this time, each female will release millions of eggs. Males will release even more sperm. Their bodies seem to smoke as the cloudy substance escapes into the water. The floating sperm and eggs will mingle as they float. If they touch each other, the eggs will be fertilized and will start to develop as they drift on the ocean currents.

## The Larval Phase

It does not take long for fertilized eggs to develop into tiny **larvae**. Called **pluteus** larvae, baby sea urchins look nothing like adults. A pluteus has a rounded head with many hard bars poking off its rear. At first glance, it looks a lot like a miniature jellyfish. Unlike real jellyfish or adult sea urchins, however, the pluteus's body has two similar sides. Animals with this body arrangement are called **bilaterally symmetrical**.

A brand-new pluteus larva drifts freely for about twelve hours. At the end of this time, the pluteus sprouts little hairs all over its body. These hairs beat rhythmically to push the pluteus through the water. The larva swims toward the ocean surface and enters the **plankton**, a floating community of microscopic plants and animals. It will stay in the plankton from 20 to 40 days. During this time the pluteus does its best to find and eat tiny plants. It also tries to avoid being eaten by other creatures. As the days go by, the little animal gets bigger and stronger.

After a few weeks in the plankton, the pluteus begins to change shape as an adult skeleton starts to grow inside its body. At the same time, larval features such as the mouth and the gut disappear. The rapidly changing pluteus sinks to the ocean floor.

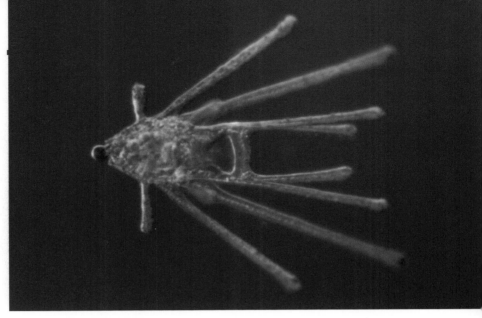

*This magnified image of a sea urchin larva shows that baby urchins look nothing like adults.*

The adult features then come out from the pluteus's body and absorb any remaining larval tissues. The entire process takes about an hour. When it is done, no traces of the pluteus remain. Measuring less than $\frac{1}{20}$ inch (1mm) across, the young sea urchin now looks just like a tiny adult urchin.

## A Different Method

A few types of sea urchins skip the pluteus stage. Instead of producing millions of tiny eggs, females of these species create a much smaller number of large, yolky eggs. They store these eggs in special pouches that line the outsides of their bodies. Sperm from male sea urchins float into the pouches and fertilize the eggs. The eggs then develop directly into young sea urchins without passing through a plankton-living phase.

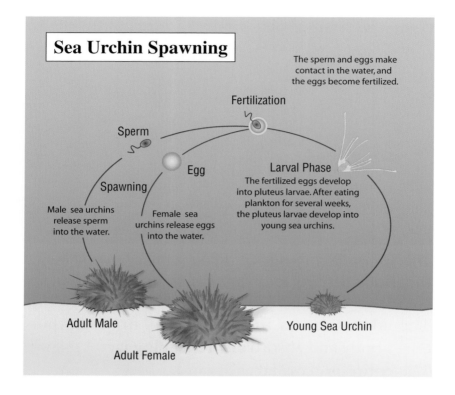

Sea urchins that reproduce in this way live mostly in polar waters. Scientists think that the harsh conditions in these areas might make it hard for floating eggs to survive. Polar species therefore have developed special ways to protect their eggs and make sure they complete their development.

## Sea Urchin Communities

Once a sea urchin is ready to start life on its own, it usually settles in a community with other sea urchins. Sea urchins often enter a community straight from the plankton, following chemical signals that tell them where other urchins live. If a

## The Sea Urchin Life Cycle

young sea urchin does settle in an unpopulated area, it may set off across the seafloor in search of other individuals.

Sea urchin communities can be enormous, containing tens of thousands of adults. The adults in a community do not really interact, but they may live very close together. Their tightly packed bodies and spines form a prickly barrier against large animals. Smaller creatures, however, can and do move freely within the community. Some little animals even make permanent homes among a sea urchin's spines. Common urchin residents include the young of certain damselfish and some worms. Small shrimp have also been found in sea urchin communities.

*These purple sea urchins live together in a tightly packed community that provides protection from predators.*

*Like all urchins, the members of this community off the coast of France are constantly on the move.*

All the sea urchins in a community move around slowly as they live and grow. As they move, they scrape away at the rocky surfaces on which they sit. Over generations the sea urchins' movement can dig hollows in even the hardest material. When many urchins live together, these hollows can stretch for enormous distances. A large hollow made in this way is called an **urchin bed**, and it may be a place where sea urchins have gathered for centuries.

Within an urchin bed, individuals may also carve out private holes. They do this by rubbing their spines and teeth against the rock below. Over time, a sea urchin can dig a hole deep enough to partly hide its body. Some species of sea urchins live in

## The Sea Urchin Life Cycle

these holes throughout their lives. When they die, new sea urchins move in to take their place.

## Growing Up

After settling down on the ocean floor, a young sea urchin begins its lifelong process of eating and growing. If conditions are good, the little urchin grows quickly. When it reaches a certain size, it is a mature adult. It is now able to spawn and produce baby sea urchins of its own. The size at which a sea urchin reaches maturity varies from species to

*Adult slate pencil urchins like this one can grow to be eight inches in diameter.*

species. The red sea urchin of California, for example, is mature when its test measures about 2 inches (5cm) across. Green sea urchins, which are found in cold northern waters around the world, reach maturity when they are about 1 inch (2.5cm) in diameter.

It is not very likely that a sea urchin will reach this point in life. Most sea urchins die within the plankton when they are still larvae. Many others are eaten by predators when they are small. Some are killed by diseases, and still others are poisoned by unclean water. However, if a sea urchin manages to reach adulthood, it has the chance to produce many, many offspring during its lifetime. With luck, some of these offspring will survive long enough to become adults and create babies of their own. In this way, the cycle of life continues, and sea urchin populations around the world remain strong and healthy.

# Chapter 3

# Eating and Defense

For sea urchins, daytime is rest time. These animals sit quietly in rocky cracks and crevices as long as the sun shines. When night falls, however, sea urchins leave their hideouts and spread out across the seafloor to search for food.

Different types of sea urchins eat different things. Most shallow-water species eat plant matter, especially algae and kelp. A few types of shallow-water urchins are **carnivores**. They eat sponges and other creatures that cannot run away. Some will even eat other sea urchins if they get the chance. Deepwater sea urchins probably eat anything they can find, including other animals (dead or alive) and plant or animal debris that rains down from above.

## On the Hunt

When looking for food, a sea urchin uses its spines and tube feet to carry itself across the ocean bottom. The urchin does not travel in any particular direction. It just moves around slowly and steadily at a top speed of a few inches (several centimeters) per minute. As it travels, the sea urchin gathers information with its sensitive tube feet. It stops for a quick snack break whenever it bumps into something it can eat.

Not all sea urchins hunt in this random manner. Some use a sense similar to smell to detect distant food. The white sea urchin of California, for example, can pick up chemicals given off by distant kelp. Once it catches this "scent," the urchin travels across the ocean floor toward its tasty seaweed meal.

Some sea urchins do not hunt at all. Instead, they sit and wait for food to come to them. These passive hunters sit in burrows and stick out their tube feet. They catch food particles with the pedicellariae at the ends of their feet or trap it between their spines. Then they transfer the food from one set of pedicellariae to the next until it reaches the mouth.

## Into the Mouth

The sea urchin's mouth, called the **Aristotle's lantern** or sometimes just the lantern, is found in the center of the test's bottom side. The mouth got its name because it was first described by the Greek nat-

Eating and Defense

*This close-up of a slate pencil urchin shows the creature's mouth at the bottom of its test.*

uralist Aristotle, who thought it looked like the top of an oil lantern. The mouth is a round hole ringed by five sharp, triangular teeth that point inward. Controlled individually by muscles, the teeth open and close like little trapdoors to seize food. When fully closed, the teeth meet in the center of the mouth to form a solid barrier.

Sea urchins that eat other animals use their lanterns for biting and tearing. Those that eat plants use their lanterns mostly for scraping plant matter off rocks and other hard surfaces. This process, called grazing, is a full-time job for a hungry sea urchin. The constant scraping wears the urchin's teeth down over time. However, the teeth grow throughout the urchin's lifetime, so they are always sharp and useful. No matter how old a sea urchin gets, it is always ready and able to find the food it needs to survive.

After the lantern seizes food, the meal passes into the sea urchin's body and moves into a small stomach. In the stomach, bacteria help break down the food. The partly digested food then travels

*A group of urchins feeds on a bed of kelp, a favorite food of shallow-water urchin species.*

through an intestinal tube that spirals around the inside of the test. By the time the food reaches the end of the tube, all its usable chemicals have been absorbed into the sea urchin's body. The remaining waste is dumped into the water through a hole in the top of the test.

## Sea Urchins as Prey

Sea urchins are not just hunters. As an important food source for many ocean animals, they are also hunted themselves. Lobsters, crabs, sea stars, seagulls, and even other urchins are just a few of the creatures that like to eat sea urchins.

Sea otters in particular are known to eat sea urchins. A hungry sea otter dives to the ocean floor and uses its flexible front paws to pick up a sea urchin and a rock. The otter then returns to the surface. Floating on its back, the otter uses the rock to crack open the sea urchin's test. Then it pries the sea urchin apart and feasts on its internal organs.

Some types of fish also have special ways of feeding on sea urchins. The parrot fish, for example, has a strong beak that can easily crush a sea urchin's spines and test. Another fish, the ocean pout, has a large head with a mouth wide enough to swallow sea urchins whole. One reef species, called the triggerfish, uses its pointed mouth to blow a jet of water at a sea urchin. The force of the water flips the urchin over, allowing the triggerfish to nibble on its prey's lower, spineless side.

*Sea otters are just one of the many predators that prey on sea urchins.*

In some parts of the world, humans are major predators of sea urchins. Urchins are hunted for their ripe gonads (called **roe**), which are considered a delicacy in some Asian cultures. In the United States, hunted species include the red, purple, and green sea urchins. More than 70 million pounds (32 million kilograms) of these creatures are re-

# Eating and Defense 35

moved from the waters of California, Maine, Oregon, Washington, and other areas each year. The roe is removed and either sold in the United States or shipped overseas.

## Defense

To protect themselves from predators, sea urchins use many defenses. The best way to avoid being eaten, of course, is to not be seen in the first place. Some sea urchins manage this by hiding during the

*A sea urchin blends in with the green coral of its surroundings to hide from predators.*

daytime and coming out only in the dark of night. Others **camouflage** themselves with colors or textures on their spines that help them blend into their surroundings. A few types of sea urchins deliberately disguise themselves by using their pedicellariae to place pebbles, shells, and seaweed on top of their bodies.

When hiding fails, a sea urchin uses its sharp spines to protect itself from predators. Sea urchins have fine control over their spines and can swivel them in any direction. When a predator approaches, some sea urchins wave their spines wildly in all directions to form a many-pointed, moving barrier. Others lock their spines into position to create a thorny suit of armor that cannot be pushed aside. Either way, a sea urchin does its best to make itself seem dangerous and unappetizing.

Despite these warnings, many predators will still attack a bristling sea urchin. When this happens, the sea urchin's spines may plunge deep into the predator's flesh. The spines go in easily, but because they are lined with backward-pointing barbs, they are very difficult to remove. Once a spine has pierced the skin, the shaft usually breaks, leaving the tip embedded in the skin. Like a giant splinter, the spine tip is hard and painful. It will eventually dissolve, but this takes many days and is very uncomfortable. An animal that has been speared by a sea urchin is not likely to try eating one again.

# Eating and Defense

As if sharp spines were not enough, a few types of sea urchins are **venomous** as well. Fire urchins, for example, have poison sacs at the tips of their spines. These sacs break open after the spines puncture a predator's skin, releasing a substance that causes weakness and sometimes death. Some other species, including sea urchins of the Toxopneustes genus, can inject venom with their pedicellariae. This defense is very effective against sea stars and other large, slow-moving predators.

*The colorful fire urchin wards off predators with its poison-tipped spines. Its venom can be deadly.*

Even nonvenomous sea urchins can use their pedicellariae to protect themselves from small creatures. The pedicellariae are particularly good for picking the larvae of other species off the body. By keeping itself clean, a sea urchin prevents these animals from feeding on and eventually killing their host.

## Nature in Balance

Despite sea urchins' many defenses, these creatures often end up as another animal's dinner. This is a good thing for the balance of nature. If sea urchins are not controlled, they can eat enough to change the underwater landscape. In the 1970s, for example, populations of green sea urchins exploded along the eastern Canadian coast for reasons that scientists have not been able to discover. The hungry urchins destroyed kelp beds hundreds of miles (hundreds of kilometers) long. The beds stayed empty until the 1980s, when a disease wiped out most of the area's sea urchin population. Only then were the kelp beds able to thrive once more.

A similar situation occurred along the west coast of North America. At one time the area's sea otters were hunted for their fur. As the sea otter population shrank, kelp beds suddenly disappeared. Studies were done to find out why. It turned out that the otters had been eating sea urchins and controlling their populations. When the sea otters started to disappear, the urchins' numbers grew and the

*In the 1970s, huge numbers of green sea urchins like these destroyed hundreds of miles of kelp along the Canadian coast.*

kelp beds were eaten. Laws were eventually passed to limit the hunting of sea otters. The sea otter population then rebounded, more sea urchins were eaten, and the kelp beds returned to a healthy state.

Although sea urchins can be destructive, they are very important to the balance of the underwater world. Without sea urchins, many ocean plants would grow out of control. Kelp beds, for example, would become thick and unhealthy. Sand flats would disappear as sea grasses sprouted wildly. Coral reefs would become unrecognizable as algae spread over them, hiding the reefs' beautiful rocky structures. In short, many habitats in today's seas could not exist without the help of sea urchins. These creatures may be simple and sometimes unremarkable. By just living and eating, however, they make the ocean world a more interesting place.

# Glossary

**Aristotle's lantern:** A sea urchin's mouth.
**bilaterally symmetrical:** Having two opposite but nearly identical sides.
**camouflage:** Coloring that helps an animal to blend into its background.
**carnivores:** Animals that eat the flesh of other animals.
**echinoderm:** The group of animals that includes sea urchins, sea stars, sea cucumbers, and sea lilies.
**Echinoidea:** The scientific class to which sea urchins belong.
**gonads:** Organs inside the sea urchin that make eggs or sperm.
**larvae:** Young sea urchins before they change into their adult form.
**macroscopic:** Visible to the naked eye.
**pedicellariae:** Tiny pinchers on the ends of some of a sea urchin's tube feet.
**plankton:** A floating community of microscopic plants and animals.
**pluteus:** A sea urchin larva.
**primary spines:** The sea urchin's longer spines.
**radially symmetrical:** Having similar parts evenly arranged around a central point.

**roe:** A sea urchin's ripe gonads.

**secondary spines:** Shorter spines that poke out between the primary spines.

**spawning:** Releasing eggs and sperm into the water.

**spines:** Long, hard objects that stick out from a sea urchin's central body.

**test:** A sea urchin's hard, rounded skeleton.

**tidal pools:** Rocky holes along the shoreline where pools of water remain during low tide.

**tube feet:** Hollow, fleshy tubes with suction-cup ends that poke out of a sea urchin's test.

**urchin bed:** A scraped-out hollow in the seafloor where many sea urchins live.

**venomous:** Able to inject poison into another creature.

# For Further Exploration

## Books and Magazines

Beth Blaxland, *Sea Stars, Sea Urchins, and Their Relatives: Echinoderms.* Philadelphia: Chelsea House, 2003. Defines echinoderms and describes the physical characteristics of many different species.

Mary M. Cerullo, *Sea Soup: Phytoplankton.* Gardiner, ME: Tilbury House, 1999. Includes fabulous photos and information regarding planktonic algae and other plants.

Sarah Lovett, *Extremely Weird Animal Defenses.* Jackson, TN: Avalon Travel 1997. Slate pencil sea urchins are just one of the unusual animals featured in this book.

Wheeler J. North, "Giant Kelp: Sequoias of the Sea," *National Geographic*, August 1972, pp. 250–69. Read about the giant kelp beds of California and their near destruction by sea urchins.

## Web Sites

**About.com**, "About Reef Tank Janitors" (www.saltaquarium.about.com/cs/aboutjanitors/a/

aa052899_3.htm). Includes sea urchin trivia and profiles of some particularly interesting sea urchin species.

**Stanford University,** "Uni's World" (www.stanford.edu/group/Urchin/uni). In this game, players move sea urchins through their environment as they eat, spawn, and avoid predators. Click on "Help" to read the rules of the game.

**Coral Reef Network**, "Sea Urchins" (www.coralreefnetwork.com/stender/marine/echinoderms/urchins/urchins.htm). Includes full-color pictures and descriptions of seventeen sea urchin species.

# Index

Aristotle's lantern, 30–33

banded urchin, 12
bilateral symmetry, 22
brain, 17
breathing, 16

camouflage, 36
Caribbean reef urchin, 12–13
carnivores, 29
collector urchin, 12
communities, 24–27

damselfish, 25
defense, 35–38
digestion, 32–33

echinoderm, 9
eggs, 8, 19, 21, 23

fire urchin, 37

gonads, 21, 34
gravity, 16–17
grazing, 32
green sea urchin, 28, 34, 38

heart urchins, 9

hunting, 30

kelp beds, 38–40

lantern, 30–33
larvae, 22–23, 28
light, 16, 29
long-spined urchin, 12

macroscopic animals, 10
magnificent urchin, 12
mouth, 30–33
movement, 16–17

nature, balance of, 38–40
night, 16, 29, 36

pedicellariae, 16, 30, 36, 37, 38
pinchers, 16
plankton, 22, 23, 24, 28
pluteus, 22–23
pollution, 8
predators, 16, 28
prey, 33–35
purple sea urchin, 34

radial symmetry, 16
red sea urchin, 18, 28, 34

reproduction, 8, 18–24
roe, 34–35

saltwater animals, 9
sand dollars, 9
sea otters, 33, 38–39
sea urchins
  communities, 24–27
  defense, 35–38
  eating, 29–33
  habitat, 10–11
  identifying, 12–13
  life cycle, 18–28
  maturity, 27–28
  movement, 16–17
  as prey, 33–35
  *see also individual species*
senses, 16
shingle urchin, 12
skeleton, 13–14
slate pencil urchin, 12
smell, sense of, 30

spawning, 19–24
sperm, 8, 19, 21, 23
spines, 6, 9, 11–13, 30, 36
spiny urchin, 12
stomach, 32

teeth, 31, 32
ten-lined urchin, 12
test, 13–14, 15, 21
Thomas's urchin, 13
tidal pools, 11
tube feet, 15–16, 30

urchin bed, 26

venom, 37

water pressure, 15–16
water quality, 6, 8
West Indian sea egg, 12
white sea urchin, 30

# picture credits

Cover photo: © Stephen Frink/CORBIS
© Brandon D. Cole/CORBIS, 10, 11, 32
© Stephen Frink/CORBIS, 13, 20
© Richard Herrmann/Visuals Unlimited, 25
© Andrew J. Martinez/Photo Researchers, Inc., 19
© Amos Nachoum/CORBIS, 7
National Geographic/Getty Images, 34, 39
Brandy Noon, 17, 24
© Photodisc, 14
© Jeffrey L. Rotman/CORBIS, 26, 31
© Science Pictures Limited/Photo Researchers, Inc., 23
© Peter Scoones/Photo Researchers, Inc., 27
© Stuart Westmorland/CORBIS, 37
© Lawson Wood/CORBIS, 15, 35

# about the author

Kris Hirschmann has written more than 100 books for children. She is the president of The Wordshop, a business that provides a variety of writing and editorial services. She holds a bachelor's degree in psychology from Dartmouth College in Hanover, New Hampshire. Hirschmann lives just outside Orlando, Florida, with her husband, Michael, and her daughters, Nikki and Erika.